Church PA System Handbook

Charlie Edwards, PhD

Certified Audio Engineer

CHURCH PA SYSTEM HANDBOOK
Published by Edwards Ministries, Inc., Chattanooga, TN

ISBN: 978-0-9914146-0-4

Other books by Charlie Edwards:
CARIBBEAN SENTINEL (novel) Tate Publishing,
BIBLE WHITE PAPERS, Edwards Ministries, Inc.
Published in the United States of America

DEDICATION

This book is dedicated to Pastors, Worship leaders, and the countless volunteers who operate the sound systems in the churches where the Gospel of Jesus Christ is preached around the world.

CONTENTS

ACKNOWLEDGMENTS

The author took the cover photo of the main audio console at Dallas Bay Baptist Church in Hixson, TN. We wish to thank the folks there for permission to use this photo on the cover of this book. Special thanks to Sound Engineer, John Cady for his help. Also we wish to express our thanks to the folks at Sweetwater.com for giving permission to use some of their definitions in our glossary section. All quotes from Scripture come from the Authorized Version.

CHAPTER 1

THE ULTIMATE PURPOSE OF THE PA SYSTEM

We live in the richest country on the globe. There is a multitude of ways in which this truth is manifested. Americans are blessed with grocery stores literally filled with every imaginable type of fresh fruits, vegetables, meats, baked goods, canned goods, etc. The streets are choked with citizens who drive cars – nice cars. Every home is air-conditioned in the summer and heated in the winter. We enjoy and take for granted the finest clothing, appliances, utilities, communication, transportation, health care, etc. – all at our fingertips. We can thank God for all of it. One more item on the list is the sound reinforcement equipment in our churches. Its purpose is to allow us to hear what goes on and be able to enjoy the worship service and ultimately take pleasure in our fellowship with Christ. Like all the other modern conveniences, this expensive equipment is designed to make life easier for us. But what happens after the guys that sold it to you and installed it leave? Someone is put in charge of this equipment and asked to operate and maintain it. How is that working? There are two answers to this question.

In many cases someone has stepped up to the plate and taken on the assignment and has performed well. Perhaps they had some previous experience of some kind that enabled them to gain some knowledge of audio. But then there is the other side of the coin. In many of our churches across the country, the PA system is the pastor's nightmare. Feedback problems haunt every service. The music never sounds "just right." You can't hear the part you wanted to hear, other parts are too loud, mysterious ghost-like echoes plague the services, annoying hums, buzzes, and pops draw attention away from the program, etc. As you gaze at all the pilot lights, indicators, meters, and switches you wonder, "What is all this stuff and why can't we make it sound like it should?"

This book was written as an aid to the pastor, staff-member, or church member who has a PA system "with problems." This book will help you resolve your audio problems and give you an understanding of what is going on with each component in your system. Our goal is to help the churches that don't have a resident trained audio expert. In most cases the people who operate the sound system are not professional audio technicians. They are truck-drivers, business owners, sales people, painters, or postmen. The vast majority has absolutely no training with audio gear. They don't understand the difference between resistance and impedance, capacitance and inductance, AC and DC, notch filters and low-pass filters, et al. Contrast this with all the professional live shows that go on in the world. Take a look at the sound technicians at the big music shows in the large arenas. They are skilled professionals and they know exactly what they're doing. Everything is executed properly. There is no feedback, ringing, pops, delays, and other bug-a-boos. Why is it that the worldly music shows are skillfully done and many Christian programs settle for mediocrity?

We are living in the last days before our Lord returns. Soon the trumpet will sound and the church will be caught up in the air. **"For the Lord himself shall descend from heaven with a shout, with the voice of the archangel, and with the trump of God: and the dead in Christ shall rise first: Then we which are alive *and* remain shall be caught up together with them in the clouds, to meet the Lord in the air: and so shall we ever be with the Lord."** (I Thes.4:16-17).

Until that happens the devil will do everything in his power to disrupt church services and cause confusion. One of his favorite things to "monkey with" is a church PA system. After all, he is the god of this world (2 Cor. 4:4) and electricity is right up his alley. Remember in Luke 10:18 the Scripture says, **"...I beheld Satan as lightning fall from heaven."** In this verse, Satan is compared to lightening. What is lightening? Lightning results when a positive static charge builds up in a mature cumulonimbus cloud. This charge reaches a high enough voltage that it seeks to neutralize itself into a negative source, whether another cloud or ground. The velocity of lightning is a much higher rate of speed than that of sound, thus it "breaks the sound barrier" and results in thunder. (Note: there are several theories on the cause of lightning and thunder, none of which are provable.) When the cloud neutralizes itself into a ground potential the result is current flow and energy. In plainer words, we have an electrical circuit. What are we saying? The Scripture relates Satan to electricity, electronics, voltage, current, etc.

Some of his work results in poorly or improperly installed audio equipment, improper operation of equipment, and untrained audio technicians. The unfortunate results are poor level control, unbalanced music performances, too loud, too soft, un-equalized music tracks, feedback, phantom oscillations, ear "squinting," and visitors who don't return. Perhaps the most important negative effect is a distracted preacher. A pastor studies all week in preparation for his sermon. When the time comes to deliver the message the audio system rears its ugly head. Everything he says rings. He thinks about that ring instead of the text and his message. A distracted preacher, an unsaved sinner who was also distracted and did not hear the Gospel, and the devil could not be happier. Again, our purpose is to help the average church resolve those nagging PA problems.

A sound system should be invisible. That does not mean that you cannot see the speakers, microphones, or mixing console. It means when you sit in the auditorium and listen to the speaker, you can hear and understand every word he says, just as if the two of you were in a private conversation. That is, no echo, reverb, ringing, pitch change, distortion, no over-emphasis of bass or high frequencies – just the natural sound of his normal voice. The invisible sound system is your ultimate goal. Few systems will deliver that, but many come close.

The reason for even having a pulpit facing multiplied rows of pews is for somebody to speak and someone to listen. Therefore, the clear communication of the voice to the audience is the goal and the purpose of spending all that money on the audio equipment. The phrase, **"He that hath an ear, let him hear what the Spirit saith unto the churches;"** is found seven times in the Scripture. Equally as important is *whom* the Spirit addresses. All seven times the Holy Spirit is addressing "the churches." God wants His message to be heard – through the ear. The phrase, **"…let him hear…"** is found seventeen times in Scripture. The word, "hear" is found 412 times in the word of God. Let's look at a few examples:

"Gather yourselves together, and hear, ye sons of Jacob; and hearken unto Israel your father." (Gen. 49:2).

"Observe and hear all these words which I command thee, that it may go well with thee, and with thy children after thee for ever, when thou doest *that which is* good and right in the sight of the LORD thy God." (Deut. 12:28).

"Jesus answered and said unto them, Go and shew John again those things which ye do hear and see:" (Matt. 11:4).

"Ye men of Israel, hear these words; Jesus of Nazareth, a man approved of God among you by miracles and wonders and signs, which God did by him in the midst of you, as ye yourselves also know:" (Acts 2:22).

"Blessed *is* he that readeth, and they that hear the words of this prophecy, and keep those things which are written therein: for the time *is* at hand." (Rev. 1:3).

In all of the 412 times "hear" is used in Scripture there are some instructive things to learn: Someone is *saying* something to someone else. The importance of that person *hearing* what is said is significant. The actions are contingent upon the *hearing* of the message.

The Scripture is full of admonitions that must first be heard before something else takes place. For example: **"And *that* their children, which have not known *any thing*, may hear, and learn to fear the LORD your God, as long as ye live in the land whither ye go over Jordan to possess it."** (Deut. 31:13). The children must first *hear* before they can learn the fear of the LORD.

Paul spoke of the importance of communicating a message clearly, **"And even things without life giving sound, whether pipe or harp, except they give a**

distinction in the sounds, how shall it be known what is piped or harped? For if the trumpet give an uncertain sound, who shall prepare himself to the battle? So likewise ye, except ye utter by the tongue words easy to be understood, how shall it be known what is spoken? for ye shall speak into the air." (I Cor. 14:7-9).

In Romans 10:14, Paul emphasizes the importance upon *hearing* the message of the preacher, **"How then shall they call on him in whom they have not believed? and how shall they believe in him of whom they have not heard? and how shall they hear without a preacher?"** The question is asked, **"How then shall they call on him…"** (to be saved) unless they *hear*? Paul sums up and gives us the reason behind the importance of *hearing* the message, **"Even as I please all *men* in all *things*, not seeking mine own profit, but the *profit* of many, that they may be saved."** (I Cor.10:33).

The word, "hear" is used 130 times in the New Testament. The Greek word used is "akouo," which is the root word for **acoustics.** Part of the definition of "akouo" is "to understand." Therefore, the importance of a well-tuned and operated sound system in a modern church auditorium is paramount in the great commission given to the Church. And thus if follows that the greater rationale for operating a problem-free sound system is not the elimination of annoyances for Christians, but rather the clear conveyance of the Gospel message to the unsaved.

CHAPTER 2

THE HISTORY OF PUBLIC ADDRESS

It is our belief that the word of God is not just a sacred book about how we got here and how to get saved, but rather The Final Authority in all matters of faith and practice, including history. That is, if the Bible says it, we believe it, take it literally (unless it is impossible to take it literally), and apply it to our daily lives. If all the history books say one thing and the Bible says something different, we go with the word of God. From Genesis through Revelation the reader of Scripture will find numerous occasions where someone is speaking to a crowd of people. Most of those crowds were quite plentiful. One would think that back in the Old Testament some of those speakers knew something about being heard by large crowds. Moses spoke to hundreds of thousands of Israelites. How did he do it?

God told Moses to go speak to the children of Israel. There were thousands of them! The Bible says in Exodus 3:14, **"And God said unto Moses, I AM THAT I AM: and he said, Thus shalt thou say unto the children of Israel, I AM hath sent me unto you."** Moses decided that the crowd was too large and worked out a way to down-size the group. **"And Moses and Aaron went and gathered together all the elders of the children of Israel:"** (Ex. 4:29). **"And Moses spake unto the heads of the tribes concerning the children of Israel, saying, This *is* the thing which the LORD hath commanded."** (Num. 30:1). Another way of communicating to large numbers of people was the use of ram's horns. We read in Joshua 6:5, **"And it shall come to pass, that when they make a long *blast* with the ram's horn, *and* when ye hear the sound of the trumpet, all the people shall shout with a great shout; and the wall of the city shall fall down flat, and the people shall ascend up every man straight before him."**

The ram's horn is the most basic of sound reinforcement systems. It is a simple megaphone. There is nowhere in Scripture that says they used it as such, however, the reference is made in a passage in Isaiah 58:1, **"Cry aloud, spare not, lift up thy voice like a trumpet, and shew my people their transgression, and the house of Jacob their sins."** The trumpet is also a basic megaphone and refers to a ram's horn in the passage listed above from Joshua. Another reference is found in Revelation 4:1, **"After this I looked, and, behold, a door *was* opened in heaven: and the first voice which I heard *was* as it were of a trumpet talking with me; which said, Come up hither, and I will shew thee things which must be hereafter."**

When we come to the New Testament, we can learn some things from Jesus Christ about addressing large groups. In the feeding of the 4000 and the 5000 Christ was in a desert area away from the city. The city is a source of noise. The wilderness is quiet. He taught the people in a quiet environment.

Another occasion, Christ found Himself in the midst of a great multitude and wanted to speak to them. The scripture says, **"And great multitudes were gathered together unto him, so that he went into a ship, and sat; and the whole multitude stood on the shore"** (Matt. 13:2). Christ stepped into a boat and pushed away from the shore and sat in the boat and taught. The people gathered around Him on the shore.

Acoustic wave propagation tends to rise as it is transmitted. Christ knew He needed to be in the lowest point and His audience above him, in order to use sound wave propagation to His benefit. The reader will also notice that Christ sat down in the boat to maximize the effect. The Greeks used this same concept in the design of their theaters. The actors were at the lowest point and the audience sat in a large rising semicircle around the stage. Thus, the speakers could address hundreds of people with great effectiveness. Properly constructed auditoriums will have a sloping floor down to the platform area with the majority of the seating *above* the podium.

The band-shell is an acoustic reflection device, which naturally transmits sound. In many cases such as the Hollywood Bowl, there is rising seating situated in front to enable an audience to hear performances without the aid of electronic equipment. Early preachers used a sounding board situated behind them to reflect sound energy forward. This device is similar to the hinged top of a grand piano, which is elevated to reflect the sound from the instrument.

Let's move ahead in time to the present day and get into the modern sound system. There is one very essential chapter which we need to address before we get down to business. In order to understand a sound system, you really need to have a handle on the basics of electricity and electronics. This certainly will not be an exhaustive study. However, a general understanding of the essentials is necessary.

CHAPTER 3

THE BUILDING BLOCKS

Immediately we hear a groan come up from the reader. "Why do I need to know all that stuff?" Good question. Any good technician starts with the basics in what ever field he pursues. How are you going to resolve an unwanted oscillation if you don't know what an oscillation is? If you see a button that says, "3 dB Pad," will you know what its function is? If you have a local radio station coming in on your PA system, how will you get rid of it? If your sound system pops every time someone turns on a certain electric light and you are asked to fix it, will you be able to? How about a buzzing microphone? Any idea why it's buzzing?

Yes, you need to know some basics all right. I can already hear somebody saying, "well so and so has been running the sound system for years and he doesn't have any electronic training." So and so would be doing a better job if he did have some basic understanding of electronics. So let's get with it and look first of all at the elements of electricity.

Voltage and Current

First, we will discuss voltage and current. The first thing I'm going to do is give you a similitude. Let's imagine a garden hose hooked up to a water spigot. For the purpose of this illustration, let's say the water that flows from the spigot through the hose is current. The water pressure will be voltage. So, a half of a turn on the water faucet is going to provide enough pressure to push the water through the hose at a trickle. Two full turns on the faucet gives a much better flow of water. Crank the handle wide open and you really have a stream of water flowing through the hose. The same is true with voltage and current. Voltage is the (electromotive) force that pushes current through a circuit. Voltage seeks ground potential. If it finds a way to get to ground, it uses its energy to get there. If it has to travel through the filament of a light bulb to get to ground, so be it.

In your home you have an electric circuit-breaker panel. It has 110 and 220 volts (plus or minus a few volts depending on where you are, relative to the power source). The appliances which draw the most current, such as clothes dryers, oven, air conditioner, etc. use the larger amount of voltage and the remainder of the electrical devices works on 110. This power source is known as Alternating Current or AC. At the power generating station there are large rotating turbines which produce positive and negative current flow during each rotation. The alternating characteristic lends itself to providing significant amounts of power over great distances of cable. These voltages are extremely high and must be reduced through substations and ultimately stepped down with a transformer to 220 volts on the pole near your home.

Resistance

Resistance is simply the opposition to the flow of current, measured in ohms. Anything in a circuit that opposes current flow is resistance. Let's go back to the water spigot and the hose. Imagine the water is turned on and you are washing your car. The hose gets twisted and results in a kink in the hose. The water flow reduces to a trickle. The kink in the hose is the resistance in the circuit and opposes the flow of water through the hose.

When you plug in an appliance into your electric socket, it provides a path for the current to ground through a resistance. Some examples of resistors in your home are: the heating elements in your toaster, coffee maker, and hair dryer. Other electrical devices which draw current are your computer, TV, and the sound system in your church auditorium. In a DC circuit, resistance can be measured with an Ohm meter. The symbol of the ohm is the Greek omega - Ω.

Resistors are used in electronic circuits for various purposes. A resistor will "drop" the voltage of the circuit depending on its value. If a particular component in the circuit operates at 6 volts and your power supply puts out 12 volts, you will need a method to reduce the voltage to the operating limits of the device. The uses for resistors are too numerous to list in this book. Resistors have color coded bands which indicate the amount of resistance and percentage of accuracy.

Capacitance

A capacitor is an electrical component, which opposes a *change* of voltage in a circuit. The capacitor is capable of storing voltage when "charged up." Capacitance is measured in farads. Capacitors, sometimes referred to as "condensers" or "caps," are used in a variety of applications.

They offer voltage regulation and filtering characteristics in electrical power supplies. They are handy in fixing audio problems also, which we will see later. Capacitors, used in tuned circuits are key elements in audio equalizing circuits. In many cases if you push the "low cut" or "bass roll off" button on an audio console, you just switched a capacitor into the circuit to filter out low frequencies. A capacitor acts as a mechanical shock absorber in an electric circuit.

Capacitors are found in power supplies and are used to regulate DC voltages. As capacitors age they may weaken and become the source for unwanted hum in an audio system. The audible hum is actually the DC output of the power supply being modulated by 60-cycle AC that was filtered by weak capacitors. Capacitors are also used in speaker crossover networks. They are used in equalizers. Capacitors will maintain their charge after an electrical device is turned off and unplugged. One must always exercise caution when working with larger capacitors in complex electronic circuits because of these characteristics.

Inductance

A simple inductor is coil of wire. When current flows through this coil there is an electromagnetic force that is generated which tends to resist a change of current in that circuit. This electromagnetic field (that is generated around the windings of the coil) results in a "shock absorber" effect on the flow of current in the circuit. Inductors are used in conjunction with capacitors in power supplies for voltage and current regulation. Also referred to as a "choke," or "coil," the inductor is used in audio circuits for equalizing sound. Much like capacitors, they are used in filtering networks as well. The device is simply a coil of wire around an iron core and sometimes an "air core." Inductance is measured in "Henries."

Inductors can be used to remove RF from audio lines in a "low pass" filter. Like capacitors, inductors are used in a multitude of applications including audio. When you were younger, you may have wrapped many turns of wire around a nail and hooked the wire to a dry cell battery to make an electro-magnet. That was an inductor you fabricated. If you put a piece of paper over the top of it with iron filings on it and watched as they formed lines around the magnet, you visualized that electromotive force that opposes the flow of current.

Reactance

When capacitors and inductors are part of an AC circuit, they maintain a constant state of change. The capacitor is charging and discharging. The electromagnetic field in the inductor is constantly reversing itself, and thereby opposing the current change. This constant state of change within these components, offers additional resistance to the flow of current in the circuit. This additional element of resistance is called Capacitance Reactance and Inductive Reactance. Reactance contributes to the Impedance of the AC circuit. This is instructive to the audio technician for the following reasons:

1. The audio in your PA system is a complex AC circuit.

2. Audio cables offer a certain amount of capacitance, proportionate to their length. The materials from which the cable is made also factor into the equation.

3. In an AC circuit this line capacitance takes the form of capacitance reactance and affects the frequency response of the circuit as well as signal level. The capacitance will tend to negate the higher frequencies, thus giving the audio a

"muddy" sound.

4. Coils of audio cable (while in a circuit) offer inductive reactance to the circuit and affect the audio in a way opposite of the capacitive reactance. With inductive reactance the lower frequencies are attenuated.

5. While the variations caused by capacitive and inductive reactance in the circuit are relatively insignificant (short cable runs), the audio tech should be aware of the potential problem.

6. For high impedance cables, physical lengths should be kept to a minimum (unbalanced line level, high impedance mikes, electric guitar cables, etc.).

Impedance

Impedance is the sum of resistance and reactance in an AC circuit. It is measured in ohms and symbolized by the letter, "Z." In audio we have high and low impedance components. This applies to microphones, low and high level amplifier inputs and outputs, as well as transmission lines. In many cases (but not always) you can tell by the type of audio connector whether it is low or high impedance. For example, an unbalanced audio line with a ¼" phone plug, RCA plug, or mini-plug will generally be a high impedance connection. A balanced line, that uses a Cannon type or XLR connector, generally is a low impedance line. While we are speaking in general terms, broadcast and PA audio uses low impedance connections (with the exception of some recording and playback gear), while consumer type audio uses high impedance. The advantage of low impedance connections are:

1. Longer cable runs without significant loss.

2. Rejection of RF.

3. Rejection of hum and other unwanted noise sources.

4. The audio connectors are generally more durable.

Impedance cannot be measured with an ohm meter. If you measure the resistance of the voice coil of an 8 ohm speaker, you will see a reading of approximately 20% less than the rated impedance. The reason for this is that reactance is nonexistent until there is AC in the circuit. An ohm meter measures with a DC voltage.

Power Supplies

All AC powered audio equipment must have a power supply in order to operate. Normally the 110 volts AC enters the component (mixer, amplifier, processor, etc.) and is rectified to a DC voltage. This task is accomplished in the power supply. The four basic parts of the power supply are: 1) power transformer, 2) diodes, 3) filter capacitors, and 4) voltage regulation. The line voltage enters the transformer primary at 110 and is "stepped down" to a lower voltage via the transformer "turns ratio." In some equipment (vacuum tube amplifiers, for example) use much higher DC voltages and must be stepped up rather than down. Turns ratio is the ratio of the number of windings in the primary to the number of windings in the secondary. If you have a transformer with 100 windings in the primary and 50 windings in the secondary, you have a turns-ratio of two to one. If you put 110 volts on the primary, you will have (disregarding losses) 55 volts on the secondary. The voltage is "stepped down" to 55 volts. When the voltage is reduced one half (as above), the current capabilities double. If you step the voltage down from 110 to 27.5 volts (four to one turns ratio), you quadruple the current capabilities. It is the current that gets the work done in the circuit.

The lower AC voltage comes out the secondary winding of the transformer and passes through a diode configuration, which rectifies the AC to an unregulated DC voltage. The use of filter capacitors in parallel with this circuit opposes the change of voltage with its "shock absorber effect." The constant charging and discharging of the capacitor bank maintains a constant DC voltage. Additional help with this task is completed with a built-in voltage regulation circuit. Now we have a clean and regulated supply of DC voltage and current to supply power to the audio component.

Amplifiers

There are hundreds of audio amplifiers integrated into a PA system. The output of a dynamic microphone represents a very low AC voltage on an incoming transmission line. This line is plugged into a mike input on the audio console and enters a low level amplifier or pre-amp. The microphone signal is amplified anywhere from 40-60 dB. The line output of the audio console is anywhere from +22 to +28 dB. In each audio channel of the mixer there are equalizers, attenuators, pan circuits, etc. Each of these adds loss to the circuit and must be compensated by additional gain. Built-in amplifiers keep the audio gain at specified levels. In addition, they offer a buffering effect. The caveat is that each amplifier stage adds to the noise level in the circuit.

Power amplifiers take the output of the audio chain and boost it to drive loud speakers. Other amplifiers drive numerous other devices including level indicators, VU meters, distribution amps, headphone amps, effects send and return circuits, et al.

Oscillators

An oscillator is an amplifier, which injects part of the output into the input. It is designed to operate in a constant state of feedback. The frequency of oscillation is determined by the values of the RC network (resistor/capacitor) within the feedback circuit. An oscillator, also called an audio generator is a handy piece of test equipment. Fixed or variable frequency (sweep) audio generators are used to test the capabilities of audio systems as well as troubleshoot them. However, oscillation can be unwanted. When the gain of your sound system is increased to a certain level, it will begin to oscillate. That is, the microphone picks up audio from the speakers, injecting output back to input. The goal is to achieve desired audio levels well below the feedback threshold.

Your local power company provides each customer with an "audio oscillator" operating at sixty cycles per second (Hertz). However, you do not want to use their "oscillator" for testing your audio system. On the contrary, sixty cycle "hum" is a nuisance in any audio system and later we will look at ways to eliminate it (levity).

Frequency Spectrum

One of the most important concepts for the audio technician to understand is the Frequency Spectrum. The normal human ear hears sounds from approximately 20 to 20,000 cycles per second. As we age, the high end drops off as well as the low. DC to 20 cycles is undetectable to the human ear. If you tuned an oscillator to 15 cycles, amplified it, and fed an exposed speaker, you could see the speaker cone moving back and forth (at a rate of 15 times per second), but would not hear the tone. These low frequencies are known as sub-audible frequencies. Constant exposure to high Sound Pressure Level (SPL) damages the human ear mechanism, thus decreasing one's ability to hear the upper and lower end of the frequency spectrum. Environments such as industrial presses, jack hammers, rock concerts, cockpits of small aircraft, stock car racetracks, shooting ranges, and other high SPL surroundings can permanently damage the human ear. When you see the kids with 500 and 1000 watt audio amplifiers in their automobiles and the trunk lid rattles with each boom, there is one thing for certain: when he reaches 40 years of age, he won't be able to hear anything below 250 cycles or above 12,000 cycles (unsubstantiated speculation on author's part).

The sound technician should be very familiar with the audio spectrum. He should know approximately what a 400-cycle tone sounds like, as well as a 1000-cycle and 10,000-cycle tone. As the frequency increases, the wavelength becomes shorter. Conversely, as the frequency decreases, the wavelength becomes longer. This concept will be extremely important later when we get into audio wave propagation in the section on audio speakers. An easy way to remember this is with radio towers. An AM radio station, on low end of the Standard Broadcast Band (with a carrier frequency of 540 kilocycles) will have a much taller tower than a station on 1540 kilocycles. The tower height represents a quarter wavelength of the frequency on which the station operates (This is a general rule. In some cases tuning networks may be incorporated to alter electrical and thus physical tower height.). The tower itself is the antenna. When we look at the FM band, we increase the frequency to a bandwidth of 88-108 megacycles. At this frequency, the wavelength is much shorter and smaller. The tower becomes the support for the antenna, which is mounted on top. In fact several antennas are stacked for antenna gain. You might see twelve "bays" stacked on top of each other at the top of the tower. If the antenna on your cell phone is one inch long, you know it is a very short wavelength, and thus a much higher frequency than the AM station operating on 540 kilocycles. Speaking of radio; generally frequencies above human hearing are referred to as radio frequency, or "RF." RF can be a nuisance to an audio system. Later we will look at ways to eliminate the RF problem, if it exists in your system.

Noise

A good sound technician should have a general understanding of noise. What is it? What causes it? How do I get rid of it? What is "signal to noise ratio"? You will have a certain amount of noise in your sound system no matter what you do. The goal is to minimize it.

The first thing we will look at is 60-cycle hum. You may have noticed by now that I use the term "cycle" instead of Hertz. Please be kind enough to cut me some slack on this issue. I'm old school and like the sound of cycles per second rather than Hertz. Okay, that nasty hum that we sometimes hear is lovingly called 60-cycle hum. It is caused by the same commercial power that operates your sound system. It's getting into the system somewhere. Your job is to find it and eliminate it or attenuate it as much as possible. By the way, remember Luke Skywalker's and Darth Vader's light sabers? When they fought with those things, did you happen to notice the sound they made? That's 60-cycle hum. So that would indicate they operate on regular AC line voltage. Right? I don't recall seeing any power cords trailing behind Luke and his father. Therefore, they must operate on batteries (DC). Right? At this time I am unable to proffer any reasonable answer on how you get 60-cycle hum from DC operated mechanism. Chalk it up as literary license.

Back to the hum in your system: First let's isolate the hum. Is it coming in on a mike cable? All audio cables should have a "shield," which is ground potential. Pin one on the cannon connector is ground and should be grounded at the audio console. Depending on the situation, the other end should be left "floating" (unattached). There can be installations where the ground is connected on both ends. The sending end of the shield sometimes is lifted to reduce hum or unwanted noise.

Hum gets in from a lack of grounding. It represents a differential in voltages between one or more components or terminations. A good ground system eliminates these voltage issues by making everything the same potential. An amplifier with aging filter capacitors can also offer hum. Sometimes isolation transformers will help eliminate hum in the circuit. Cable placement sometimes will contribute. If you have a low-level audio source near a power cable, electromagnetic force will affect the audio in the form of hum. A good rule is to keep your power cabling away from audio cables (if possible).

Sometimes lighting will contribute to hum in a system. The ballast unit in florescent lights are bad about radiating spurious emissions that may be induced into audio cables, and thus generate hum. In many cases mike placement will fix this problem. In other cases an impedance mismatch might be the culprit. If there are two or more audio devices connected in series, resulting in unwanted hum, check the levels of each device. An oscilloscope is an excellent piece of test equipment to track down hum problems.

Signal-to-noise ratio is just that. It is a ratio of your desired program audio level to that of the background noise of the system. It is desirable to have a considerable spread between the two levels. Let's illustrate a bad signal-to-noise ratio. Many of us have heard poorly recorded programs on local "dollar-a-holler" Gospel radio stations. In many cases the preacher is physically too far away from the microphone. Therefore the microphone picks up the air-conditioner, the airplane going over the building, and a dog barking at the same level as the voice. Another illustration: When that same radio preacher sets the record level on an audio recording device too low, very little modulation is indicated on the VU meter. This will also result in poor "S/N" ratio.

Frequency Response

Frequency Response is a subject that should be understood by any good sound tech. Each component of your sound system has a frequency response. It is the measure of an audio system's output as a constant-level input frequency changes. Ideally you want the same output level at all audio frequencies. In plainer words, if you inject a 50 cycle tone at an amplifier's input (at a given level), and the tone is measured at –3 dB at the output, your component has a 3 dB loss at 50 cycles. Normally, the entire audio spectrum is measured at the output with audio frequencies inserted at the input. The measurements are given relative to a certain level. For example, "20 to 20,000 cycles plus or minus .5 dB." This example shows an amplifier which has the ability to amplify the entire audio frequency spectrum at the same level, within a half a decibel of the zero level. This would be considered a relatively "flat" frequency response and would also be considered a favorable one. An example of a poor frequency response with the same frequency spectrum might be, "20 to 20,000 cycles plus or minus 3 dB." The level drop will most likely be at either or both ends of the spectrum and would not be "flat," but curved. The device being measured might be a low-level amplifier, line amplifier, buffer amplifier, power amplifier, microphone, headphones, loud speaker, etc. Ideally, the audio component will replicate the same amplitude throughout the spectrum (flat). There are times, however, that a high-end boost, or "pre-emphasis" is built into a circuit. This means the high end of the frequency response would be measured at a higher level than the lower end. This is a common characteristic built into a lavaliere microphone. Because of the position of the mike (on a tie-clip and not close to the mouth), the higher frequencies tend to be attenuated and the voice sounds "tubby." With a built-in boost on the high end, this characteristic tends to be corrected and the voice sounds more normal with proper levels on the sibilant frequencies.

This low and high equalization of the frequency response shows up on basic audio systems as "bass" and "treble" knobs. This is a basic form of equalization.

Level

Audio comes in a variety of levels. The output of a microphone is typically minus 50 dB (decibel). The output of the audio console is typically plus 10-20 dB. When an audio signal passes through an equalizing device, such as an RC network, losses are incurred. We have seen how the capacitance reactance adds impedance (resistance) to the circuit, which, in turn brings the audio level down. The level must be boosted back to operating level with an amplifier. Conversely, a signal which might be at too high a level must be attenuated down to a more tolerable level (by your system). This might show up on your audio console as a "mike/line" switch, "input level" knob, or a "pad."

Ideally, all inputs on the audio console will be at a level which will allow the operator to run the individual pot (volume knob or slide fader) at the zero level indication on the scale. On most audio consoles with slide faders, this position is approximately two thirds of the distance of the slide from the bottom. It is always marked in some form. If you are running a mike wide open, you need to boost the signal. If you are operating it close to the bottom of the slide, you should attenuate the signal in order to operate at the zero indication.

Most audio gear will have a VU indicator of some sort. Most no longer use the analog VU meter, which utilizes the D'Arsonval meter movement, but have switched to LED indicators. However, the volume unit or "VU" is still measured. The scale on this meter is from –20 to +3 dB. If you go down to your local radio station and look at the same meter, it will indicate a little different. At the zero point, it will read 100. At the –20 point, it will read zero. The difference relates to the modulation level of the signal. For the purpose of PA systems, we will stick to the first scale mentioned, showing 0 dB at two thirds of the scale.

CHAPTER 4

THE EQUIPMENT

Audio Console

The audio console is the main control point of the system. It is also probably the most misunderstood of all the components in the system. Every operator should familiarize himself with the console equipment manual and learn all its nuances. It is at this point in the system where the operator can adjust and tweak the audio to the desired point. The operator's ears are critical to his operation of the console. If his ear detects a flaw in the audio such as an unwanted low frequency emphasis, this should be corrected using the console. If he sees that another audio source is too "hot" and pegs the needles or gets a "clipping" indication from an LED peak indicator, this should be corrected with input attenuation. Every program element is different and must be adjusted to optimum. This will be covered in greater detail in the section on equalization.

The console can be compared to a musical instrument. When a talented musician sits down at the piano, the result is beautiful music. Similarly, the sound system in the hands of a skilled operator has the same results.

An audio console has inputs and outputs – and lots of them. The console should be sized according to the need of the individual church. The size of the console has much to do with the number of inputs and outputs. Each individual input will be either a microphone or line-level audio source. A typical dynamic microphone has an output of a minus 50-60 dB and must be boosted to a line-level via a first-stage mike preamp. The output of a CD player, for example, operates at line-level and does not need boosting. Other music sources or program elements operate at various levels and need to be adjusted with an input level control so that the main slide-fader or "pot" in the "zero db" position (12 O'clock on a rotary pot and about 60% on a slide pot) will show a 1000 Hz tone at zero on the VU meter.

Each individual input channel has a number of different controls. We mentioned the input attenuator. The number of these controls depends on the complexity of the console. Typically, each channel has a "pan" control, which will allow the operator to adjust that audio source to the output left and right in varying degrees. Each channel may also have fixed output selector. This fixes that source to a particular output channel. Individual channels will also have an "effects send" pot, which allows the controller to adjust the level of audio sent to an outboard audio processing device. The output of this device is returned via the "effects return" jack and injected back into the circuit. Each channel will also have equalization available. Each channel (depending again on complexity) may have its own visual audio level indicator. Many individual channels have a "roll-off" button, which filters out anything below a specified frequency. There may also be a control for phantom power supply for microphones which operate on a DC voltage. Most modern consoles have a mute button for each channel. The operator should learn the correct use for each of these features and utilize them to optimize the audio.

Riding Gain

The term, "riding gain" simply refers to the operation of the audio console. There are some rules of good practice that should be noted.

1. **The cardinal rule in operating the sound is, never have anything open that is not in use. That is, the only open mike or channel on the board is the one being used now. If the mike is not being used, the pot or fader should be turned to the infinity (off) position or the mute button activated.** Each time the number of open microphones is doubled, the gain must be lowered 3 db in order to avoid feedback.

2. Always test a new microphone set up. That is, if a group is to sing and there are multiple mikes used, a simple level check on each one can save a headache later.

3. Always know what is going on in terms of order of events. In other words, who speaks next on which microphone? What changes have to be made to the board to accommodate a smooth and unnoticed change in the sound system?

4. Avoid turning equipment on and off during operation. In many cases (depending on how things are connected) an audible "thump" is heard over the PA system because the voltage spike produced in the process gets amplified through the system.

Microphones

Microphones come in all shapes and sizes. There are various types of mikes used for different applications. Typically, a dynamic microphone is used for voice and is connected to the sound system through a cable and a plug. This mike operates by a transducer that moves a coil back and forth in a permanent magnetic field in response to an acoustic source such as a voice. A small AC voltage is generated by this action, which is just the opposite of an audio loudspeaker. The small voltage is then fed into a mike preamp.

The condenser microphone operates with the aid of the phantom power supply mentioned above. This microphone uses the two plates of a capacitor powered with the voltage from the console (or power supply). One of the capacitor plates is a diaphragm, which vibrates with the audio source. This vibration changes the physical distance between the two plates and thus changes the value of the capacitance. The slight change of capacitance causes the voltage to fluctuate above and below the bias level. This voltage fluctuation is detected across a series resistor and amplified for an audio signal.

There are several types of microphones available. However the two mentioned above are the most common. The dynamic and condenser microphones are utilized in different ways. The wireless microphones can be either dynamic or condenser types. Normally, a wireless mike, which is attached to the lapel or necktie of the speaker, has a built-in pre-emphasis at the high end.

Microphones are designed with different polar patterns in mind also. A directional microphone will have the highest output when the microphone is pointed directly to the acoustic source. An omni-directional mike will pick up audio at the same level in all directions around the mike element. The directional microphones are available in varying cardioid patterns. There are super-cardioid and hyper-cardioid patterns as well as "shotgun" mikes according the need of the application. All this should be kept in mind for the PA operator to get the sound he wants while at the same time minimize noise. You would not want to use a super cardioid microphone as a single mike for a quartet. It would be better if each member had one. (Don't forget to take the time to explain to each quartet member about keeping the mike pointed at the voice. While you're at it, tell them not to pump the mike in and out of range for some imagined effect. You should have equipment in the rack to take care of that.) There are also available bidirectional microphones, which offer pickup in opposite directions. This would be acceptable in situation where dialog between two voices is miked. For hanging mikes in the choir, use omni-directional. For mikes on the platform, use a standard cardioid mike.

Normally, microphones are low impedance. The definition of impedance is the total opposition (in an AC circuit) to the flow of current. Impedance takes into consideration all the inductive, capacitance, and miscellaneous resistances and reactances found in an AC circuit. Impedance is measured in ohms. Typically, an audio speaker has an impedance of eight ohms. A microphone has an impedance of anywhere from 50 to 200 ohms, all of which are considered low. The output of most CD players and other "hi-fi" equipment is considered high impedance. An electric guitar is a high impedance audio source.

There are high impedance microphones, but they are generally not used in PA systems or for recording. Their use is limited due to the need for a short cable. At higher impedances, the line loss is greater. Low impedance microphones are better suited to the needs of PA systems and broadcast studios, where long cables are used between the mike and the audio console.

For church PA use, the best all around microphone to use is the low impedance dynamic mike. The wireless application is good for a speaker to be free of a podium and have the ability to walk around. The down-side of the wireless mike is the constant need to turn it on and turn it off and change the batteries. The simple dynamic mike with its cable and cannon connector cannot be improved upon.

There are a few things to remember about microphones. Encourage those that use the mikes to never blow into a microphone. There are other ways to test a mike to make sure it is working. Blowing on it tends to stretch the transducer membrane and can damage it. NEVER check a microphone with an ohm meter. That will destroy the element. You can check the mike cable, but make certain to remove the mike from the circuit. NEVER install a condenser mike with a phantom power supply anywhere near a baptistry. It is possible (this has happened) for someone standing in water to touch the mike and be electrocuted. Or if somehow the mike touches the water when someone is in the water, fatal electric shock can occur.

Amplifiers

A public address system is made up of many audio amplifiers. They can start with a small chip inside a microphone and work their way through the system to the final power amps that drive the speakers. We mentioned the mike preamp earlier. Every component in the system that routes, switches, equalizes, changes level, modifies the audio in any way offers loss in the system. The loss is recovered in the amplifiers. The goal is keeping the signal amplified and the noise level down.

Head-room in each of the amplifiers is also a key to quality sound. You will want plenty of power in reserve so when there is a crescendo in the music or program signal, it gets amplified as needed without clipping (distortion). This rule is thrown out when it comes to guitar amplifiers. Most of the amps available come with hundreds of different settings to obtain various types of modified, over-driven, and clipped wave forms. Just the opposite is desired with the PA system. An abundance of headroom in the amplifiers will assure plenty of dynamic range in the system. Remember, the ideal sound system is invisible. That means it doesn't sound like a sound system – it sounds real.

The low-level amps are for getting the signal up to line-level. The intermediate amps operate at line-level and keep the signal in that range, buffering the various stages. Finally the power amps take the signal and drive the speakers.

Impedance matching is vital between amplifiers. The output of one stage must operate comfortably with the input of the next. An example might be a long transmission line plugged into a line level input at the console. There are times when an audio transformer is needed to match two different elements. In many cases, the audio device uses a built-in transformer for this purpose. Transformers are also used for isolation of long transmission lines such as phone lines. They are also a valuable tool for matching balanced and unbalanced audio elements.

Power amplifiers are the final stage of the system. If there are volume controls on a power amp, they are typically opened wide and volume is controlled through earlier stages. The output of this amp has a much lower output impedance than those of earlier stages. You will see anywhere between 4 – 16 ohm outputs on power amps. Adding and subtracting speakers will alter the impedance of the speaker load, depending on how they are added. For example, by adding an

8-ohm speaker to an existing 8-ohm speaker (in parallel) it will change the load impedance from 8-ohms to 4-ohms. If the speakers are wired in series, the output load changes to 16 ohms.

When the load impedance changes from 8 to 16 ohms, the characteristics of the power amplifier output stage change. With the additional impedance, the current flow is lowered. The formula for power is P (watts) equals current (amps) squared times resistance (ohms). So with a 100 watt amp operating into an 8 ohm speaker (at 100 watts), the formula is 100 = current squared x 8 or P x R divided by current squared. Current = 3.53 amps. When you change the resistance to 16 ohms, the current changes to 2.5 amps. Less current means less volume.

Speakers

Loudspeakers are the final stage of the sound system. The speaker is a transducer that changes electrical energy into acoustic energy. A typical speaker utilizes a conical-shaped diaphragm, which freely moves back and forth within the bounds of a basket frame. At the base of the cone is the voice coil, which is housed in the electro-magnetic field of a magnet. As the voice coil is modulated with the output voltage of the power amp, the reaction of the opposing magnetic field causes the speaker cone to move in and out and radiate sound waves in the air.

A quality speaker will have a relatively “flat” frequency response. The term, “flat” means it reproduces all frequencies in the audio range with the same amplitude. It is difficult for any one speaker to accomplish this. Therefore, speaker systems are assembled utilizing different speakers which are designed to replicate specific areas of the audio spectrum. Crossover networks are used to split the signal into two or more paths to feed the audio signal to each speaker. By using high and low pass filters; a crossover network can channel frequencies above or below a certain frequency to a particular speaker or additional network.

Each speaker must be wired in phase with other speakers in the system. Speakers wired out of phase tend to cancel each other out. The terminals on speakers and speaker cabinets are marked for this purpose.

Speakers designed for lower frequencies are called woofers or subwoofers. A typical crossover frequency would be 800 Hz. So, a woofer would handle all frequencies between 20 and 800 Hz. The lower frequencies tend to have different wave propagation than the higher end of the band. This phenomenon is noteworthy when it comes to speaker placement. Just about anywhere you put a woofer or low frequency driver, it will be heard perfectly anywhere in the listening area. Low audio frequencies will find their way to the listener. The opposite is true for the higher frequency speakers and drivers. They are more line-of-sight or directional in nature. Normally in a speaker array, there will be one or two low frequency cabinets pointed in the general direction of the audience and several high frequency horns pointed at all areas of the room. Each horn tends to splay the higher frequencies out in a predetermined field. In many cases, this field is rated horizontally and vertically. For example, a typical horn may have a horizontal field of 100 degrees and a vertical field of 60 degrees.

Speaker Placement

When choosing a position to place the speakers in a sound system, there is much to consider. Usually in the case of a temporary set-up the best location is on each side of the stage or platform. However, in the permanent installation a split system is not always the best. The optimum for speaker placement is a single point. Therefore, in many church auditoriums the speakers are hung from the ceiling over the center platform rather than on each side. The main benefit of placing the speaker array in a single point is intelligibility. We will say more about that in a subsequent chapter.

Most sound systems have multiple outputs according to the needs of the church. The main speakers are referred to as "House." Speakers that point back to the platform are referred to as "Fold-back" or "Monitor" speakers. There are times when additional monitor speakers and/or headphone jacks are needed, depending mostly on the needs of the music program.

Equalizers

Equalizers are the sound man's best friend. One of the objectives of a sound system is to provide plenty of audio gain without oscillation (feedback). In order to accomplish gain without feedback, the sound system must be equalized. Every room or auditorium has its own "acoustic signature" and no two are the same. There will be at least one resonant frequency with harmonics. Harmonics are simply multiples of a particular frequency. For example: if you play the middle "A" key on the piano and measure it with a frequency counter, it will show 440 Cycles. The second harmonic of 440 is 880 Cycles. The third harmonic is 1320 Cycles and so forth.

In an unequalized sound system there will be one frequency that resonates above all the others. It's easy to find. Simple bring up the gain until the system goes into oscillation and there is your resonant frequency. It is the job of the equalizer to compensate for these unwanted peaks in the frequency response of the system. Ideally, a mirror image of the system frequency response scheduled into the room equalizer should produce an overall "flat" response for the room. How is this accomplished?

A Real Time Analyzer (RTA) is used to present a visual display of your system's frequency response. There are models available with onboard equalization. You will want one equipped with generators for white and/or pink noise. Many also provide the more traditional audio sine-wave generator for a specific frequency. Pink noise is the most common source signal used to equalize a sound system. What is the difference between white noise and pink noise? White noise is an amalgam of all audio frequencies heard at approximately the same level across the audible frequency band. If you listen to an FM broadcast station and detune the station, you'll hear white noise. If the station goes off the air while you're listening, you'll hear it. It has a "hissing" characteristic to its overall sound. Pink noise is essentially the same thing with a slight difference. Rather than all frequencies reproduced at the same level, pink noise increases in amplitude as the frequency decreases. There is more amplitude for lower frequencies. A graphic display of this signal would appear as a downhill slope from left to right. The resulting sound is very much like the sound of a waterfall or the pounding surf at the beach and could be described as roar. Interestingly, pink noise is mentioned in Scripture: Psm. 93:4, **"The LORD on high *is* mightier than the noise of many waters, *yea, than* the mighty waves of the sea."**

The RTA equalizer feeds pink noise into the audio console and "listens" to the sound system with its own microphone and analyzes the signal. It will digitally equalize the system "on the fly." For example: Your sound system is in operation before your church service begins (perhaps recorded music plays during this time) and remains in service throughout. People steadily file into the auditorium and fill in vacant seating. As more and more people populate the building, the acoustic characteristics of the room change. The

RTA equalizer automatically compensates and maintains a flat output for your sound system. It is certainly worth the relatively small amount of money it costs.

Many RTA equalizers also come with a Sound Pressure Level (SPL) indicator. This is another valuable addition to your system. This will enable you to maintain a comfortable SPL for your auditorium. Nothing is worse than sitting in an auditorium and regret not bringing your foam ear plugs because the audio is simply too loud. This can actually occur if the sound technician is too far away from the speakers. In many cases the audio console is up in the back of the balcony and the technician is operating at appropriate levels for that position. The audience up front, closer to the speakers, may be enduring uncomfortable SPLs and wincing at the preacher's every verbal emphasis. OSHA's occupational noise exposure limits are listed on their web site, www.osha.gov. For a period of one and one-half hours (typical church service) the sound pressure level is not to exceed 102 db. Obviously these numbers are for the work environment and not a worship service, however, every church sound technician needs to be aware of any SPLs straying from the comfort zone.

25 or 70 Volt System

Most church systems have additional areas (apart from auditorium) where sound is desired. Those areas might include offices, nursery, vestibule, rest rooms, hallways, etc. The most efficient way to provide sound for these areas is with a 25 or 70 volt audio system. With the use of a constant voltage amplifier, multiple speakers can be driven from a single amp. Each speaker can have its own volume and on/off control. This type of system lends itself to long lengths of cable between speakers.

These speakers are usually found mounted in ceiling panels. The speaker itself has a small transformer attached with a number of taps available, depending on how much volume is needed in that area. The relatively low impedance of the speaker (8 ohms) is isolated with the use of the transformer. The speaker is connected to the secondary of the transformer, with the primary connected to the 25 or 70 volt source.

Each transformer will have a number of primary taps. Each one represents an amount of audio power in RMS watts that is applied to the speaker. If you have 10 speakers in the system and you use the 5 watt tap, you would want at least a 50 watt amplifier. The total number of watts on the transformer taps used must be equal to or less than the RMS power output of the amplifier. It is also very important to use shielded audio cable with the shield grounded only at the amplifier end of the circuit. The speaker end of the shield is not connected. For long lengths of cable running through attics and above ceiling panels will invariably run near noisy florescent lighting and AC cables. Without a grounded shield around the audio cable, stray voltages could be induced into the lines and cause problems. I was called in to repair an installation one time where someone had used unshielded CAT 5 cable (designed for Ethernet) to wire the speakers in a 70 volt system. This cable was run in parallel bundles with several AC power lines. Those AC lines induced over 180 volts of AC (as read on a volt meter) onto that CAT 5 cable. Needless to say the audio contained distortion and caused problems with the audio amplifier. Shielded audio cable grounded properly eliminates this problem.

Miscellaneous Equipment

Depending on what type of ministry you have, you may want some additional equipment in your audio rack. A distribution amplifier is a handy item to have for most applications where you need additional audio feeds. The "DA" has a single input and multiple outputs. There may be a need to send audio to a phone line or STL for broadcast purposes. The 25 or 70 volt system mentioned above could be fed from the distribution amplifier. Another output from the DA might feed an audio recorder or pod-cast feed. All these could be fed from the multiple outputs of the distribution amplifier).

Depending on the music that is heard in your auditorium, you will want some audio processing equipment available. There is broad scope of various and sundry processes available from a simple reverb to echo, phase shift, flanging, chorus, octave enhancement, et al. Again, this will depend on the type of music used.

Cassette (still in use) and CD players are needed in many cases where audio tracks are played for vocalists. There are CD players that have "auto-stop" feature so that the next track does not automatically play. This feature is found on karaoke equipment as well as the DJ-type CD players. Most of these machines also have an adjustment for pitch control, which could be a useful addition.

Most churches record sermons and music regularly. There are several methods of recording audio from a computerized audio system hard-drive, CD recorder, as well as the traditional reel to reel machine (still in use) and cassette decks.

Some test equipment is helpful to have around at times. A simple VOM (Volt Ohm Meter) for checking resistance, continuity, and voltage is worth having. As mentioned earlier, an oscilloscope is a good tool to have when needed. There are cable checking devices and signal injectors and detectors, which make trouble-shooting easier.

CHAPTER 5

THE CABLE

There are two types of transmission lines used in audio. It is not uncommon to find both used in any system. The characteristics and proper uses for each of these is useful information for the sound tech.

Unbalanced Line

This type of transmission line consists of a single conductor with a grounded shield and is most common in consumer audio equipment. The audio cables used to interconnect audio components with "phono plugs" or "RCA plugs" utilize an unbalanced line. The shield is a chassis ground and the center conductor is the "hot" side. These lines are normally found in pairs with a red and white plug to separate the left and right side of a stereo pair.

Another common use for the unbalanced line is for instrument cables – particularly for electric guitars, basses, keyboard instruments, and instrument amplifiers. These instrument cables commonly utilize a quarter-inch "phone plug" connector.

There are some microphones which use this type of cable. These mikes are not suited for the modern audio system due to the limited cable lengths and high impedance. However, some of these early microphone designs are popular with harmonica players because of the distortion characteristics of the mike design. In many cases the harmonica player will have his own instrument amplifier which also enhances the desired sound.

The unbalanced line is a high impedance line with a nominal impedance of 50,000 ohms. This type of transmission line is limited in its use because of the high impedance characteristics. It is susceptible to RF and other ambient noise elements. Cable lengths of more than ten feet can result in line capacitance, which alters the frequency response considerably. This is more evident in the audio component hookup cables. These cables should be kept to a minimum length to maintain audio quality.

Balanced Line

The balanced line is used extensively in sound reinforcement and broadcast use. This line consists of a pair of wires surrounded with a grounded shield. It is considered to be a low impedance line of approximately 600 ohms. The long-reaching cables needed for PA systems require this type of transmission line. Its characteristics also include better frequency response over long lengths as well as excellent noise rejection. The most common audio connector used is the "Cannon" or "XLR" connector. This is the desired type of cable used for most of the PA system cabling.

It is possible to convert balanced to unbalanced and vice-versa. For example if you have a low-impedance microphone cable that needs to be connected to an unbalanced quarter-inch phone plug input: simply solder the hot side of the pair to the center conductor of the plug and combine the low side and shield to the ground side of the plug.

For converting an unbalanced to a balanced line, an audio transformer is needed. The hot side of the unbalanced line is connected to the high side of the primary, with the low side connected to the ground. Both sides of the secondary become the two balanced line elements with the transformer ground connected to the shield. This is essentially what a "Direct Box" is (in its passive state).

Direct Box

In the early days of studio recording a microphone was placed in front of a guitar amplifier for pickup of the audio. While this is still widely done today, another option is available for transferring the guitar signal into the recording or public address system. The Direct Box allows a direct connection between the guitar and the audio console while maintaining all the appropriate impedance and level matches. It also allows a direct connection from the output of a guitar pedal board or guitar amplifier sans speakers to the audio console or similar device. The Direct Box can be active or passive with a variety of typical features such as ground lift, impedance matching, balanced to unbalanced or vice versa, gain, buffering, as well as phantom power supply.

Snake

A snake is a long cable made up of many smaller shielded audio cables that connect an audio console to microphones and monitoring equipment on a stage or platform. The ends of the cables typically will use balanced audio connectors such as Cannon/XLR connectors. In many church sound system installations a Snake cable will connect the audio console to the platform.

CHAPTER 6

THE COMMON PROBLEMS

Feedback

One of the most common sound system problems is feedback. This occurs when the output of the speaker system is picked up by a microphone and oscillation results. This is always annoying for those in the listening area of the system. Normally, a single frequency will make itself heard as microphone gain increases. A gradual feedback may begin with someone speaking into a microphone and a "ringing" will be heard as he speaks. At other times, depending on how the audio console is set up the system might go immediately into an ear-piercing howl as soon as a channel is opened by the operator. In many cases the operator has difficulty achieving sufficient gain in the system and continues sliding or rotating the fader until the feedback howl is introduced.

Over the years I have noticed that high frequency feedback affects people in different ways. What might be tolerable, but a nuisance to some can be a painful and extremely uncomfortable audible experience for others. With that in mind, the audio technician should always try to avoid feedback – especially the high pitched variety.

Every room has a different size, shape, and volume. Each room also has a Resonant Frequency. The audio spectrum spans from 20 to 20,000 Hz. depending on individual human frequency response. We will concern ourselves with frequencies within the human audio range for this book even though a room (that contains a PA system) may have subsonic as well as ultrasonic resonant frequencies. While a room may have a variety of resonant frequencies, there will be one that stands out among all the rest. This will be the first one you hear in the sound system.

Fortunately there are ways to fix the unruly and annoying resonance we call feedback. The sound system needs to be tuned to accommodate the acoustics of the auditorium. A common fix to this problem is the Room Equalizer. This device divides the audio spectrum up into segmented frequency bands, each with a fader. With the system set to an acoustically "flat" condition on the audio console, steadily increase the gain with a single microphone until the howl begins. Select the fader closest to the resonant frequency and decrease that fader, thus "notching" that frequency out. This method allows the system more volume throughout the audio spectrum except on the resonant frequency.

This is a simplified version of tuning a room. There may be harmonics of the resonant frequency that must be dealt with. Additionally, there is a variety of equalizing equipment on the market today to take care of a feedback problem. In many cases a simple solution to a feedback problem is speaker or microphone placement. If a microphone picks up a speaker there will be feedback. Make certain that the speakers are mounted in front of rather than behind the microphone.

Another simple fix is a common sound system operator error. The rule of thumb is: never have a mike opened unless you need it. Many operators leave several mikes open when they are not being used. Another mistake is the idea that two microphones are better than one for the same source. WRONG! Not only will you get phase cancellation using more than one mike (unless they are on separate output channels as left and right), but it also contributes to feedback. Remember: For each time you double the number of open mikes on a system, the overall system gain (before feedback) is reduced by 3 dB. Or, if you have a stationary pulpit mike open in addition to a wireless mike on the preacher's lapel that is turned on, your system gain is reduced by 3 dB compared to using a single mike. A gain of 3 dB is equivalent to doubling your power. A loss of 3 dB is equivalent to dividing your power by 2. **Always turn off all microphones not in use.** (I cannot stress this rule enough. This is one of the most ignored rules in church audio.)

Intelligibility Issues

Intelligibility is simply the clarity or lucidity of your sound system. In other words, when someone speaks through the sound system while you sit in the auditorium, can you understand what he says? This factor becomes extremely challenging when it comes to some installations. It is one thing to have an expensive audio system installed with all the whistles and bells. It's another thing to be able to understand every word spoken through the system from all sections of the room. Why?

Sound travels at approximately 1100 feet per second.

So, if you are sitting 110 feet from the speaker system, it will take 100 milliseconds (1 second equals 1000 milliseconds) for the sound to reach your ears. If (at that distance) you are sitting under the balcony and there happens to be a ceiling speaker 11 feet over your head, it will take 10 milliseconds for *that* sound to reach your ears. So, you hear the sound and then hear it again 90 milliseconds later. Since the main speakers are mounted on each side of the platform, there is yet another speaker which radiates from a point 175 feet from your ears. That sound takes 159 milliseconds to reach your ears. So, you are hearing the same sound three times within a single second. When you add unwanted echoes in auditoriums without any sound absorption materials, this only exacerbates the problem. With each additional repetition, the intelligibility decreases.

Ideally, you want a single sound source to reach your ears. This maximizes the intelligibility. If possible, a single main speaker array is better than two for this reason. Decreasing the number of times per second that you hear the same sound is the goal of optimizing a sound system. Sometimes the auditorium needs physical changes made to minimize reverberation. There are different methods used to acoustically treat a room to eliminate echo. Sound diffusers such as baffles, slat resonators, bass traps, anechoic panels, drapery, or changing the shape or angle of a wall will be enough to make an audible difference. In the final analysis it all contributes to the simple need for an audience to be able to clearly understand speech through the system.

One method of increasing intelligibility in a system is the use of a delay system. This device is inserted in the audio chain between the output of the audio console and the amplifiers for the ceiling speakers under the balcony. The distances are measured and a delay time is extrapolated and dialed into the unit. The result is: when you sit under the balcony, the sound from the speaker 11 feet over your head reaches your ears at the same time you hear the main speakers. This, in turn increases intelligibility.

Another method to increase intelligibility is to add little boost on the sibilant frequencies (2000 Hz to 10,000 Hz). Discretion is needed here. Other factors come into play such as the system equalization. If the un-equalized sound system tends to dip at these frequencies a little boost is welcome. However, the opposite is excessive sibilance, which is also a problem. "De-essing" equipment is available for this problem.

I attended a graduation ceremony some years back, which took place in a large cavernous church auditorium. This was a new building with nice carpet and comfortable pews, but there was one major problem: you couldn't understand what was being said over the PA system. I noticed as I sat there that I could understand about half to a third of what the speaker was saying and that was if I worked hard at listening. After a few minutes I realized what the problem was. The sound was bouncing off of a multitude of surfaces before it got to my ears. So I was hearing echoes of echoes of echoes, etc. Each syllable of speech was repeated many times as it bounced around the auditorium and reached my ear milliseconds after the one before. The level of each recurrent arrival was a little different than the one before. The room needed some serious acoustic treatment.

Noise

Another irritating problem in a sound system is noise. Noise comes in all shapes and sizes. Sometimes it's a subtle hum underlying everything that goes through the system. Sometimes it manifests itself in a gritty snapping sound heard. Sometimes you may hear a pop in the system for some unknown reason. Maybe you can hear a radio station coming in somewhere in the system; the level is way down, but it is there. These nuisances can contribute to the gray hair of any pastor. For each of the types of noise there is a reason for its existence. They can be fixed.

One of the noises mentioned is hum. This could be introduced into the system by a number of reasons. It could be a filter capacitor in the power supply of an audio amplifier somewhere in the audio chain. As amplifiers age, it's usually the large capacitors that show age first. When they lose their ability to filter the AC it shows up as an alternating fluctuation in the DC supply. This fluctuation occurs at a rate of 60 cycles per second, or what we call hum. So, one fix would be to change out the filter caps in the power amps.

Another cause for hum might be a ground issue. Perhaps there is a balanced audio pair with a shield that is improperly connected. Only one end of the shield should be grounded and that is the "destination" end of the cable. It might be a piece of equipment in the rack is not properly grounded. You can take a volt meter and start checking voltage differences between components. Sometimes lifting a ground somewhere will fix the problem. Too many grounds may cause a ground loop. Sometimes everything is grounded, but what are they grounded to? Make certain that you are using a proper electrical ground as your source. Don't rely on a water pipe to be a good source even though sometimes it is. Sometimes a hum can be caused by improperly matching unbalanced to balanced lines. In many cases the use of an isolation transformer or DI box will eliminate hum. It may be that someone used non-shielded cable for audio and the cable is in close proximity to an AC circuit. The AC circuit may induce alternating current (60 Hz.) into the unprotected audio cable. All audio cable should be shielded.

The gritty snapping sound could be fluorescent lighting near an unshielded audio cable. Another cause could be some inductive device in operation nearby. Inductive AC devices would include electric drills and other power tools using AC, ballasts in fluorescent lighting, electric fans, hair dryers, etc.

Years ago in the church I attended, the pastor had someone turn colored lights on and off over the baptistry for the effect of night turning to day when someone was baptized. Each time the man would switch the light, a pop was heard in the sound system. I offered to fix the problem for him. Of course he was all in favor of that. I simply put a .01 microfarad 400 volt capacitor across the light switch. Remember, the capacitor opposes a change of voltage. Each time the switch was thrown there was a massive spike put on the house circuit that came with a drastic change of voltage from zero to 120 or vice versa. The capacitor acted as a shock absorber across the connection and eliminated the spike, which eliminated the audible pop heard through the PA system. I knew it would work because I had already done that in the radio station where I worked.

The pop might be heard each time the air conditioner unit comes on or off. Sometimes it may require a larger amount of capacitance and sometimes it may be wise to let an electrician install the caps.

What about the radio station coming in over the system? It is way down in volume, but it is just loud enough to detect. The problem is RF (Radio Frequency) getting into the sound system somewhere and subsequently detected by a system component. There are a few ways to eliminate these nuisances. You can obtain an RF filter which is available at Radio Shack. Essentially these are made up of capacitors (of small values) and chokes (inductors) placed in the AC circuit. Also available are ferrite collars, beads, or donut-shaped filters. These will eliminate the unwanted radio station intrusion from your system.

CHAPTER 7

THE SOUND TECHNICIAN

This person is a key element in the entire ministry of a church. It is during the church services that the most important things take place that your church does. Preaching, teaching, and music to edify the Body of Christ all take place during your church service. One of the most important elements in this mix is the person operating the audio console. Now would be a good time to go back to the front of this book and review the introduction. When the audio isn't right the message that needs to get into the ears of the congregation is interrupted. Noise, feedback, distortion, ringing, etc. are all little annoyances that ultimately lead to someone NOT getting the message. This is the sound tech's domain, or his area of management. He is a vital part of the ministry.

The following "Parable of the Sound Techs" is a fictitious comparison between two types of sound technicians. Its purpose will be obvious. Throughout Scripture, the Holy Spirit uses this same method to demonstrate and teach truth. Hos.12:10 – **"I have also**

spoken by the prophets, and I have multiplied visions, and used similitudes, by the ministry of the prophets."

The Good Sound Tech

This person is a cheerful, optimistic type, always with a ready smile. You rarely see him with a neck time, much less dressed up in a suit. He has excellent hearing. He understands his responsibility to help make life easy for the Pastor and Music Director as well as anyone that speaks into a microphone. He acknowledges the fact that there is always something new to learn from others. He is always helpful. He is dependable, open-minded, and always seeking to learn more about audio and thus would cherish a copy of this book. He sees his job as a ministry for the Lord. He is usually a tech-savvy person who probably knows about computers, networking, and possibly some similar technical interests such as Amateur Radio, music, aviation, photography, etc. Above all, this sound tech understands the idea set forth in chapter 1 of this book. He will take it upon himself to seek out others in the congregation with similar ideals and interests and recruit them to help with the audio. He will teach the new guy how to do everything to make sure that there is always a trained person on hand to work the audio. He is there for choir practice or any other event that takes place in the auditorium. He will be there for all the special meetings, rehearsals, etc.

The Bad Sound Tech

This person doesn't smile very often and tends to grumble a lot. He is not prone to let other people operate the sound system. The people that are allowed to operate it must abide by his rulebook. He is never interested in learning anything from anyone else about audio since he already knows everything. If someone (that does perhaps have a better grasp of audio engineering) offers helpful suggestions about the operation of "his" sound system, he will come back with various phrases such as:

1. "We've never done it that way before."

2. "Well, the way we do it is a little different..."

3. "This is the way we've always done it…"

4. "I understand what you're saying, but..."

This sound tech does not look at his job as a ministry to others, nor does he understand the truths laid out in chapter 1 of this book.

CHAPTER 8

THE OPERATION OF THE SYSTEM

Control Room

The Control Room is the control point for the sound system. Its physical placement is crucial in a proper sound system. Typically, the control room is in the back of the auditorium or up in the balcony. This may be the optimal location when it comes to equipment security, comfort of the sound techs (Allows them to unnoticeably disconnect from the service, fellowship with each other, talk on phone, play with iPad or smart phone, surf internet, consume large quantities of coffee and donuts, etc. I know this from personal experience.), or ease of ingress and egress. However, as far as the most advantageous operation of the sound system goes, it may not be the best location. If the operator is up in the back of the auditorium, he cannot hear what the system sounds like down on the floor where the majority of the people are sitting. There may be a hum or noise that is unheard from that distance.

I remember visiting a church in South Carolina on a Wednesday evening: I was sitting on one side about middle of the way back. I noticed something peculiar about the audio and it took me a few minutes to figure out what it was. I could hear something that told me at least part of the system was operating, but I had to work at understanding what the pastor said. When I first entered the sanctuary I noticed that the audio console was back up in the balcony all the way back. Finally, I realized what was wrong. The fold-back or platform monitors were the only thing operating. The house speakers were not on. The pastor could hear himself in the monitor but did not notice his people laboring to understand what he said. The man riding gain didn't have a clue about the problem since he had his own monitor speaker, which obviously worked just fine. After the service I asked a man if the PA system always sounded like it did tonight. He told me that normally it did not sound like it did that night.

Now, if the sound control point had been down on the main floor in the middle of the auditorium and the operator had been listening to the house speakers, he would have known what was going on and taken care of the problem. The use of a monitor speaker positioned over the audio console in the back of the auditorium is not a problem. When there is no service going on and the operator needs to listen to something he needs it. However that speaker should not be what he listens to during a church service for obvious reasons.

I have been in church services where the volume was extremely loud. The SPL app on my phone showed pegged at 100 dB, but I knew it wasn't accurate. The sound man was up in the balcony way in the back and couldn't know how loud it was where we sat. I wondered if anyone else's ears were overloaded. It makes a difference where the audio console is positioned so that the operator can adjust the gain accordingly.

Simplicity

A sound system can be simple or complex. If a church has a supply of technical people who are readily available to learn how to operate sophisticated equipment and be faithful over the years to operate it diligently, that church is blessed. It is very possible that a smaller church does not have someone like that. If you have a state-of-the-art digital sound system installed, you might just have a problem trying to find someone that can operate it. Sometimes it is difficult to figure out how to turn the thing on. A pastor (or in-charge person for getting quotes on a new sound system) needs to be aware of overzealous audio sales reps who want to sell you more than you really need. They won't be around on a rainy Wednesday evening when all your sound techs are ill and your volunteers are standing there looking at the complex audio system scratching their heads. KIS (Keep It Simple) is an excellent philosophy to have when it comes to your audio system. Do you really need a system with half-a-dozen secret passwords (known only to a few) that only someone with a degree in Networking can operate?

Troubleshooting

In any system there will eventually be problems that appear. It is necessary that the person operating the system has an idea of the "big picture" when it comes to the audio system. For example: A vocalist on the platform is testing a microphone and there is no audible sound. The pot is open and you have no visual indication on the audio console VU meter that a signal is present. Immediately you know that you have lost the signal somewhere and the church service is about to begin. Everything else is working okay and is heard through the system. What is going on?

Each audio system should have a block diagram

depicting each component and where it is in the audio chain. The sound system guy in your church probably has an idea of how your system is put together, but is there a drawing posted or on file somewhere? Let's say there is a problem like the one in the previous paragraph and we take a look at the block diagram of the sound system: On the diagram we see that there is an outboard sub-mixer in the audio rack that feeds a digital echo processor in the audio chain. That circuit feeds back to the console through an auxiliary line in. And that particular microphone is in this sub-system. Now we take a look at the digital echo unit installed in the rack and discover that someone turned the power switch off. Turn it back on and the problem is resolved. The block diagram saved the day. Yes, the regular sound man probably would have caught that without the aid of the drawing, but what about the less experienced stand-by sound tech?

Another suggestion is to have a file cabinet in the control room where all the equipment manuals are kept. Operating instructions for each piece of equipment needs to be readily available. It would be a good idea for the most experienced sound tech write up operating instructions for the system overall and have them on hand in the event of a problem when he is not on site. These instructions should be written in an understandable manner, typed up, printed, and either posted or placed in the control room file cabinet.

This book was not written to serve as an exhaustive reference book on sound systems and troubleshooting guide, but rather a simple handbook for ministers and layman who work with audio in the church. It is therefore recommended that the reader secure books that are written specifically for troubleshooting. You will be surprised how often you can take care of problems that arise when you have a general understanding of the items covered in this relatively short book. For example: You have a hum coming from a particular microphone. The hum disappears when you unplug it. Open the XLR connector and examine the solder connections or check the cable to see if it has sustained physical damage to the shield. You will find that most of the problems that you encounter will be simple issues like the ones we mentioned

Summary

The audio systems in our churches are great tools in the ministry of our Lord. They enable us to hear the clear message of the Gospel of Christ and enjoy music that edifies our souls, lifts our spirits, as well as glorifies Jesus Christ. What we "hear" is vital because of what we read in Rev 3:22, **"He that hath an ear, let him hear what the Spirit saith unto the churches."** When we realize that the church PA system is a tool or medium through which hear what the Holy Spirit has for us, then it clearly has great importance. It is not something that should be approached "on the cheap." It deserves high placement on our church priority list. It deserves to be tuned properly and operated with people that have the correct attitude and understanding of its value. We pray that this book will serve as a help to Pastors and sound technicians in the use and operation of Church Public Address Systems.

GLOSSARY

60 Cycle Hum - A low frequency noise commonly heard in PA systems, which can result for various reasons.

70 Volt Speaker System – A type of PA system using a number of parallel-wired, transformer/speakers which works well for large areas and long lengths of cable.

AC – Alternating Current

Acoustic Absorption – The ability of a material to dampen sound waves and contribute to the elimination of echo.

Active Component – Any device in the audio chain that requires power to operate.

AGC – Automatic Gain Control - A method of maintaining a constant output level on an audio circuit.

Ambient Noise – The background noise in an environment.

Amplifier – An electronic component used to increase the amplitude of audio, voltage, or current.

Amplitude Modulation (AM) – A method of modulating a radio frequency carrier by audio modulation of the carrier amplitude.

Attenuator – A resistive device to reduce a signal level and maintain impedance matching.

Audio Frequency Spectrum – The frequency spectrum from roughly 20 to 20,000 Hz. based on human hearing range.

Audio Processing – To change the natural characteristics of an audio signal with use of AGC, limiters, equalizers, etc.

Audio Transformer – A passive device used in audio for impedance matching, isolation, RF elimination, and other uses.

Balanced Line – An audio cable consisting of a pair of wires surrounded with a shield used to ground out induced voltages and RF.

Bandpass Filter – An active or passive network of resistance, capacitance, or inductance, placed in the audio chain to allow a particular band of frequencies to pass through.

Bandwidth – A measurement in Hertz for the analog width of a spectrum of frequencies. For digital components it is the rate of data flow.

Bass Trap – An acoustic panel placed in an audio environment to offer a dampening effect to lower audio frequencies.

Bi-directional Microphone – A microphone which has a symmetrical firure-8 shaped polar response with the region of maximum attenuation at 90 degrees off axis.

Cannon Connector – A common audio cable connector for balanced low impedance lines; also referred to as XLR connector.

Capacitor – A passive component used in numerous electronic functions. It has the ability to store electrical energy and dissipate the energy when needed. It is made up of two conductors separated by a dialectric insulating material.

Capacitive Reactance – Opposition to the flow of current in an AC circuit due to the properties of capacitance in the circuit.

Cardioid Microphone – A uni-directional microphone which has a polar response curve like a cardioid. The maximum attenuation is at the rear and gradually decreases toward the axis point.

Clipping – The result of an audio signal when it is overdriven or driven past the power amplifier capabilities. The result is distortion.

Compression Amplifier – An active component in an audio chain where a relatively constant output level is needed. The amplifier reacts with attenuation to an extreme increase in signal level to avoid clipping.

Condenser Microphone – A microphone that uses a DC voltage across a capacitor to produce an audio signal. One of the plates of the capacitor is a diaphragm. As acoustic energy moves the diaphragm the voltage is modulated, thus producing the output signal. The mike requires an outside or phantom source of DC voltage.

Crossover Network – A network of inductors, capacitors, and resistors used to divide an audio signal into upper and lower frequencies, normally found in speaker cabinets. This network divides frequencies into lower, midrange, and high frequencies for appropriate transducers.

Current – The flow of electrons in an electric circuit. It can be compared to water flowing through a garden hose, where the hose represents the circuit and the force of the spigot (drive) is voltage which pushes the current through the circuit.

DA – Distribution Amp – provides multiple outputs of a single audio source.

dB - Decibel - A logarithmic unit used to define the intensity of an audio signal. An increase of 3 dB is equal to doubling the power.

DC – Direct Current; as opposed to Alternating Current, DC is a continuous voltage.

Decay Time – AKA T60 - With respect to echo and reverberation, the time it takes for audio to decay by 60 dB.

De-esser – A device that attenuates or eliminates sibilant sounds in recording or reproducing the human voice.

Distortion – An unwanted or wanted (depending on the application) change in an audio signal.

Doppler Effect – This is an audio phenomena that is heard when an audio source is in motion relative to the listener. To use a common experience, let's say you hear the sound of an aircraft approaching. The sound of the engine will increase in pitch (frequency) as it approaches. Then as it passes overhead and moves away, the pitch reverses and goes down. While the pitch of the engine never changes, it sounds like it changes because of the speed of the aircraft relative to the speed of sound at your point of listening.

Dummy Load – A resistive component that absorbs all the output power of a transmitter or amplifier to simulate working conditions for purposes of testing.

Dynamic Microphone – A device to transfer the acoustic energy of sound into electrical energy by utilizing a diaphragm and magnetic induction.

Electret Microphone – A relatively inexpensive type of condenser microphone that uses battery power.

Equalized Line – A telephone loop between two points which maintains the same audio quality at the output as the input.

Equalizer – Based on the root word, equal, an equalizer is an audio device whose function is to equal out the tonal characteristics of a sound. At least that was the idea back in the days when they were first conceived as a tool used to get flat response in telephone lines and to make up for the deficiencies in audio equipment and acoustic spaces. Nowadays it could more aptly be named an "unequalizer" since they are more often used creatively to alter the relative balance of frequencies to produce desired tonal characteristics in sounds. An equalizer has the ability to boost and/or cut the energy (amplitude) in specified frequency ranges by employing one or more filter circuits. There are many different types of EQ's in use today in many widely varying applications, but they fundamentally all do the same thing. (Sweetwater.com)

Fader – Another name for variable attenuator, volume control, or potentiometer. A fader works just like a standard potentiometer, only instead of rotating, it slides along a straight path. Faders are most commonly used on mixing boards and graphic equalizers, where it is nice to have both an easy way to move the level up and down, and to provide a sort of graphic representation of the relative levels of many channels (or frequencies in the case of an EQ). Faders have also historically been used on some synthesizers as controllers for various parameters. The name comes from the phrase "fade out." Once faders came into existence it became much easier for an engineer to do a smooth fade out. (Sweetwater.com)

Feedback – The howling sound or oscillation present when the output of an amplifier is feeding the input and a closed loop is completed.

Flat – A frequency response which is generally the same over a given band width, which for audio would be 20 to 20,000 Hz.

Fold back – Another term for a monitor system for the stage area.

Frequency – The number of times an alternating current goes through its complete cycle per second.

Frequency Modulation (FM) – A method of modulating a radio frequency carrier by alternating the frequency of the carrier.

Frequency Response – A rating of how effectively a circuit or device transmits the different frequencies applied to it.

Gain – Volume

Ground Loop – In an audio system, a ground loop can result when there are multiple components (active and passive) wired together that ideally are the same potential. With more than one power supply it is possible to have differing voltage potentials between these components. This can result if there exists more than one ground point for multiple grounds. For example a piece of equipment may be grounded to the rack in which it is mounted. Another component may be grounded to another source that isn't connected electrically to the rack. There could be a difference in potential of 5 volts between the two grounds, which could induce a hum in the system. More severe differences could be caused by incorrect wiring and lead to voltage differences of up to 50 volts or more. These differences in ground or voltage potentials, in many cases are the cause for hum in a system. In the early days, before a ground pin on electrical plugs was standard, you could play an electric guitar and get shocked if your mouth touched the microphone on the PA system (I know how this feels). The use of a common ground and correct electrical wiring can eliminate these issues.

Harmonic – A frequency that is an integral multiple of a fundamental frequency.

Hertz – (Hz.) Unit of measure for frequency (Cycles per second).

Impedance Matching Transformer – A transformer used to match two different impedances; for example, a high impedance unbalanced line to a low impedance balanced line.

High Input or Output Impedance – The rated impedance across the input or output of an audio component ranging from 10,000 to 100,000 ohms.

Impedance – The total amount of resistance to the flow of current in an AC circuit, including DC resistance, inductance reactance, and capacitance reactance.

Impedance Matching – This is the correct practice for connecting audio components in a sound system and will result in the maximum transfer of signal and audio fidelity. For example: the 8 ohm output of a headset jack is not a good source to feed the high impedance (50K ohm) input of a recording device. In this case the use of a transformer to match the impedances would result in a much better sounding signal.

Inductance – The electromagnetic effect produced by current flowing through a coil of wire, which has the tendency to oppose any change of current.

Inductive Reactance – Opposition to the flow of current in an AC circuit due to the properties of inductance in the circuit.

Inductor – A coil of wire which, when current flows through it, generates an electromotive force which tends to oppose a change of current.

Intelligibility – The ability to be clearly understood.

Intermodulation Distortion – the undesired interaction of electronic signals of different frequencies transmitted within a nonlinear system, resulting in distortion.

Kilohertz – (KHz) Unit of frequency equivalent to 1000 Hz.

Limiter – An amplifier circuit in which the amplitude of the output is prevented from exceeding a given value. Thus, it can be used to remove any amplitude variations in a signal while leaving an FM signal intact.

Line Level – A power level for an audio signal from approximately -10 dB to +20 dB into 600 ohms.

Line Loss – A gradual decrease in the quality and strength of a signal as the length of the cable increases.

Loop – A term used for a special phone line for broadcast use.

Low Impedance – 8 to 600 ohms

Masking Noise – Often found in large offices with multiple personnel; the use of sound (for example pink noise) to generate an ambient noise level for the purpose of covering up phone conversations and masking other audible information.

Microphone Boom – A mike stand used to extend the mike horizontally or diagonally.

Microphone Level Input – Approximately -50 to -60 dB (much lower than line level)

Microphone Mixer – A unit consisting of mike preamps, switches, and potentiometers (pots), used to blend and switch mikes and other program elements.

Microphone Preamp – An amplifier used to boost mike output (-60 dB) to line level which is from approximately -10 to +20 dB.

Middle C – The note "C" that is found approximately in the middle of a full sized piano. It is commonly tuned to 261.1 Hz. The pitch is represented in written music as the note on the 1st ledger line below the treble staff, or the note on the 1st ledger line above the bass clef. (Sweetwater.com)

Monitor Speaker – An audio transducer used in a control room which monitors program material.

Noise – Any sound or combination of sounds that is unwanted in an audio system.

Notch Filter – An electronic network, passive or active, utilizing resistance, capacitance, and/or inductance that tends to reduce the amplitude of a particular frequency; often used to eliminate feedback in conjunction with a room equalizer.

Off Axis Rejection – A term used regarding mike response; having to do with the amount of attenuation or noise rejection that a microphone has on both sides and behind (off axis).

Ohm – Unit of measure for resistance and impedance named after German physicist Georg Simon Ohm.

Omni-directional mike – A microphone which picks up the same amount of sound from all sides.

Oscillator – An amplifier with a feedback circuit that induces oscillation at a particular frequency.

Pad – A device inserted into a circuit to introduce transmission loss or attenuation and maintain impedance matching.

Pan – To adjust an audio signal between left, center, and right, as well as front to rear.

Parallel – A method of connecting electronic devices or components for operation, so that all of the hot leads are connected and all of the low sides are connected; as opposed to "Series."

Passive – A device that does not require power to operate.

Patch Panel – Also referred to as a patch bay or jackfield, a panel that houses multiple jacks or cable connections. Patch panels can be used for audio, video, telephone connections, and data transfers.

Peak Limiter – An active audio component that limits the output of an audio signal to prevent overloading the input of the subsequent audio component in the chain.

Phantom Power Supply – A supplier of a DC voltage used to power condenser microphones. This can be anywhere from 5 to 48 volts. In many cases this power supply is built in to the audio console.

Phase – Important to remember regarding wiring pairs of audio cables; Problems can arise if a pair is out of phase with the others.

Phase Cancellation – Phase describes where in its cycle a periodic waveform is at any given time. The relationship in time of two or more waveforms with the same or harmonically related periods gives us a measurement of their phase difference. Phase cancellation occurs when two signals of the same frequency are out of phase with each other resulting in a net reduction in the overall level of the combined signal. If two identical signals are 100% or 180 degrees out of phase they will completely cancel one another if combined. When similar complex signals (such as the left and right channel of a stereo music program) are combined phase cancellation will cause some frequencies to be cut, while others may end up boosted. Phase and phase difference is a real-world issue in areas such as electrical wiring of audio equipment, signal path, and microphone placement during the recording process. Phase reversal can be a serious compromise of sound quality or a special effect affecting the perceived spaciousness of the sound depending on the context of its occurrence. (Sweetwater.com)

Phone Jack – A ¼" female receptacle into which a ¼" phone plug is connected; used for unbalanced audio including headphones, speakers, patching, etc.

Pink Noise – An audio signal made up of random and spurious audio frequencies with a gradual increase in amplitude with decrease in frequency. The boost at the low end gives it a "roaring" sound similar to a waterfall or ocean waves rather than the "hiss" sound of white noise, which has a flat frequency response. Pink noise is commonly used to equalize a PA system.

Pitch – Another term for frequency. For example 440 Hz. is equivalent to middle "A" (pitch) on a piano.

Polarity – The proper orienting or arrangement of wire pairs with respect to hot side and ground or hot and low plus ground.

Polar Response – Graphically, a diagram which shows the magnitude of a quantity in relation to direction. Or, the characteristics of a microphone, for example, having to do with front to back ratio, front to side, etc.

Pop Filter – (Wind screen) a filter placed over a microphone to remove the over-modulated "P" sound and "B" sound from speech; usually made from Styrofoam or cloth and many cases color coded.

Pot – Short for potentiometer or variable resistor used as a volume control.

Pre-emphasis Kick – A gradual increase in amplitude with an increase in frequency (towards the upper end of the audio spectrum) for the purpose of improved signal to noise ratios and clarity. This concept has been used in the shaping of the frequency response of lavaliere microphones, broadcasting, recording, et al.

Presence –In movie-making and television production, presence (or room tone) is the "silence" recorded at a location or space with the absence of dialogue.

RC Network – A passive array of resistors and capacitors used to pass, block, or change the amplitude of a particular frequency or band of frequencies.

Real Time Analyzer – (RTA) is a professional audio device that measures and displays the frequency spectrum of an audio signal; a spectrum analyzer that works in real time. An RTA can range from a small hand-held device to a rack-mounted hardware unit to software running on a laptop.

Resistance – The measure of the amount of opposition a device offers to the flow of current. The unit of measure is the ohm. Resistance in a DC circuit is measured with an ohm meter.

Resonance –This is the propensity of a system to oscillate at a specific frequency with a greater amplitude than other frequencies. In many cases there is a "family" of resonant frequencies (often a fundamental frequency and its harmonics) which turn out to be incremental when measured. The fundamental frequency will always carry the greater amplitude.

Reverberation – The remainder of sound that exists in a room after the source of the sound has stopped is called reverberation, sometimes mistakenly called echo (which is an entirely different sounding phenomenon). We've all heard it when doing something like clapping our hands (or bouncing a basketball) in a large enclosed space (like a gym). All rooms have some reverberation, even though we may not always notice it as such. The characteristics of the reverberation are a big part of the subjective quality of the sound of any room in which we are located. (Sweetwater.com)

RF – Radio Frequency –A band of frequencies which include part of the audio spectrum as its low end and a high frequency that merges with the infrared frequencies (9 KHz to 300 GHz). These frequencies can affect church PA systems and can be radiated by the following sources: AM and FM broadcast stations, CB radios, VHF and UHF TV stations, Ham radios, wireless microphones, garage door openers, baby monitors, air traffic control, alarm systems, cordless telephones, microwave ovens, et al.

RF Filter – Essentially, a low-pass filter to allow audio frequencies, but block everything above that.

RMS – Root Mean Square; true power reading.

Roll Off (or Rolloff) – Specifically roll off refers to the action of a specific type of filter; one designed to roll off frequencies above or below a certain point. It is called roll off because the process is gradual. Hi pass and low pass filters both roll off frequencies outside of their range, but they don't immediately eliminate all frequencies outside their range. The sound is gently (or not so gently) "rolled off" with frequencies further above or below the cutoff frequency becoming more and more attenuated. Roll off steepness is generally stated in dB per Octave, with higher numbers indicating a steeper filter. 24 dB/Octave is steeper than 12 dB/Octave. (Sweetwater.com)

RTA – see Real Time Analyzer.

Rumbling – Unwanted noise from motorized equipment, including motor noise, HVAC, et al, which finds its way into the audio circuit.

Series – A wiring protocol where devices are connected sequentially as opposed to in parallel with one another. Specifically this means that each device is an integral component in the circuit path, such that if one device fails the circuit will become open and no longer function. For example, a circuit breaker is typically wired in series with the hot wire going to each circuit in a house. When the breaker opens the circuit no longer has power. (Sweetwater.com)

Signal to Noise Ratio – The name defines itself. In an audio system the audio signal should be at a much higher level than the intrinsic noise level of the system. The two levels are compared with each other in the S/N ratio. The higher the signal level is to the noise level; the better the audio quality will be. For example: if you don't set a microphone volume control high enough and compensate by raising the master board output, you will decrease the signal to noise ratio and thus increase the noise level. The mike pot should be moving the VU indicator up to the 100% level to eliminate this problem.

Snake – A cable assembly used in audio containing multiple audio cables. Typically it would have up to 50 or more individual shielded pairs all contained in one protective jacket. Normally one would see the cables terminated with XLR connectors or quarter-inch phone plugs. A snake cable is normally used to connect microphones and monitoring equipment to an audio console located away from the audio source area.

Sounding Board – A rigid surface placed above and/or behind a pulpit or speaking platform for the purpose of projecting the voice of the speaker. An illustration of this is the top of a grand piano when it is propped up to deflect the sound out to the audience.

Spectrum Analyzer – An instrument which gives a graphic display of the audio spectrum. In audio PA systems it is used in conjunction with an injected signal such as pink noise to adjust the sound system for maximum gain before feedback as well as room equalization. The display will indicate the overall frequency response of the room being equalized.

Spike – An abrupt transient that forms part of a pulse, but is of greater amplitude than the average value of the pulse. It is often heard in the speaker as a pop or thump.

SPL – Sound Pressure Level

STL – Studio Transmitter Link; either microwave, fiber, IP, or phone line.

Sub-audible Frequency – A frequency too low to be heard by the human ear (20-20K Hz.).

Talkback Mike – A microphone at the audio console for the purpose of the sound tech speaking to people in a recording studio or sound stage area.

Transducer – A device that transforms electrical energy into acoustic energy or vice versa.

Tweeter – An audio loudspeaker designed to reproduce high audio frequencies in the approximate range of 2000 Hz to 20,000 Hz.

Unbalanced Line – An audio line consisting of a single conductor with a shield. This type of line is used in consumer audio equipment utilizing RCA, phone plug, and mini-plug connectors.

Uni-directional Microphone – A cardioid microphone which exhibits a polar response in the shape of a heart. It has its highest noise rejection at the rear of the mike.

Voice Frequency –This is a frequency range that consists of those frequencies which are generally used by the human voice. Specifically, this band is from approximately 300 Hz to 3400 Hz.

Voltage – The electromotive force that moves current through an electrical conductor. Voltage produces either DC or AC.

VU Meter – Abbreviation for Volume Unit meter. A voltmeter used to measure audio voltage.

Wave Length – The distance between points of corresponding phase of two consecutive cycles.

Wind Screen – AKA Pop Filter – Used over a microphone to filter out excessive over modulation of "P" and "B" sounds in speech.

White Noise – An audio signal made up of random and spurious audio frequencies with a flat frequency response. You can sample white noise by listening to the "hiss" sound when tuning an FM radio between channels.

Woofer – A loudspeaker driver designed to produce low frequency sounds, typically from around 40 hertz up to about a kilohertz or higher.

XLR – See Cannon connector

A WORD ABOUT THE AUTHOR

Charlie Edwards has worked for a number of radio stations from Kentucky to Tennessee to Florida. Most of these stations he was involved in engineering. He also has designed and installed sound systems in churches as well as radio station control and production rooms.

He completed an Electronic Communications training program and graduated at the top of his class as well as earned an FCC First Class Radio Telephone Operator License in the early 70s. He is a member of the Society of Broadcast Engineers and holds five certifications including, Certified Audio Engineer, Certified Broadcast Radio Engineer, Certified Broadcast Television Engineer, Certified Television Operator, and Certified Broadcast Technician. He graduated with a BA from Tennessee Temple University. He also earned a Master of Theology, Doctor of Theology, and most recently he finished up the requirements for a PhD in Clinical Christian Counseling.

He has written a novel entitled, "Caribbean Sentinel," published by Tate Publishing and available at Amazon.com. He recently completed the re-write and updating project for a booklet published in 1983, entitled, "The Preacher's Radio Ministry Handbook." He has also written "Israel, God's Peculiar Treasure" and "Bible White Papers."

It is our hope that this book will be a help to pastors, worship leaders, and sound technicians in an effort to optimize their sound systems and get the clear message of Christ out to the world.

www.ingramcontent.com/pod-product-compliance
Lightning Source LLC
Chambersburg PA
CBHW071627040426
42452CB00009B/1515

* 9 7 8 0 9 9 1 4 1 4 6 0 4 *